U.S. Department
of Transportation
**National Highway
Traffic Safety
Administration**

www.nhtsa.gov

DOT HS 811 486

June 2011

Crash Prevention Effectiveness of Light-Vehicle Electronic Stability Control: An Update of the 2007 NHTSA Evaluation

DISCLAIMER

This publication is distributed by the U.S. Department of Transportation and the National Highway Traffic Safety Administration in the interest of information exchange. The opinions, findings and conclusions expressed in this publication are those of the authors and not necessarily those of the Department of Transportation, the National Highway Traffic Safety Administration, or the Federal Motor Carrier Safety Administration. The United States Government assumes no liability for its content or use thereof. If trade or manufacturer's names or products are mentioned, it is because they are considered essential to the object of the publication and should not be construed as an endorsement. The United States Government does not endorse products or manufacturers.

Technical Report Documentation Page

1. Report No. DOT HS 811 486	2. Government Accession No.	3. Recipient's Catalog No.
4. Title and Subtitle Crash Prevention Effectiveness of Light-Vehicle Electronic Stability Control: An Update of the 2007 NHTSA Evaluation		5. Report Date June 2011
		6. Performing Organization Code
7. Author Robert Sivinski		8. Performing Organization Report No.
9. Performing Organization Name and Address Evaluation Division; National Center for Statistics and Analysis National Highway Traffic Safety Administration Washington, DC 20590		10. Work Unit No. (TRAIS)
		11. Contract or Grant No.
12. Sponsoring Agency Name and Address National Highway Traffic Safety Administration 1200 New Jersey Avenue, SE Washington, DC 20590		13. Type of Report and Period Covered NHTSA Technical Report
		14. Sponsoring Agency Code
15. Supplementary Notes		

16. Abstract

Statistical analyses based on FARS and NASS CDS data from 1997 to 2009 found that a vehicle equipped with electronic stability control (ESC) had a smaller likelihood of being involved in a crash than a similar vehicle without ESC. This analysis estimates the magnitude of that reduction for different types of crashes and for different types of vehicles. Overall, ESC was associated with a 5-percent decrease in the likelihood that a passenger car would be involved in any police-reported crash and a 23-percent reduction in the probability that a passenger car would be involved in a fatal crash. For light trucks and vans (LTVs), the reductions are 7 percent and 20 percent respectively. Each of these reductions is statistically significant except the 5 percent overall effect for passenger cars. Fatal first-event rollovers are reduced by 56 percent in passenger cars and by 74 percent in LTVs. Fatal impacts with fixed objects are reduced by 47 percent in passenger cars and 45 percent in LTVs. These reductions are statistically significant.

17. Key Words NHTSA; FARS; NASS; ESC; effectiveness; fatality reduction; injury reduction; evaluation; statistical analysis; benefits; electronic stability control	18. Distribution Statement Document is available to the public from the National Technical Information Service www.ntis.gov		
19. Security Classif. (Of this report) Unclassified	20. Security Classif. (Of this page) Unclassified	21. No. of Pages 29	22. Price

Form DOT F 1700.7 (8-72)

i

Table of Contents

Executive Summary

When a vehicle is equipped with electronic stability control (ESC), it has a smaller likelihood of being involved in a crash than a similar vehicle without ESC. This analysis estimates the magnitude of that reduction for different types of crashes and for different types of vehicles. Overall, ESC was associated with a 6-percent decrease in the likelihood that a vehicle would be involved in any police-reported crash, and an 18-percent reduction in the probability that a vehicle would be involved in a fatal crash. For passenger cars, the reductions are 5 percent and 23 percent, respectively; for light trucks and vans (LTVs), 7 percent and 20 percent.[1] Each of these reductions is statistically significant except the 5-percent overall effect in passenger cars.

Estimates of ESC effectiveness in preventing crash involvement vary by crash type, but for crashes that typically involve loss of control such as rollovers and side impacts with fixed objects, these estimates are large and statistically significant. The effects are similar for fatal and nonfatal crashes and for passenger cars and LTVs. As specified by Federal Motor Vehicle Safety Standard No. 126, by September 1, 2011, all new cars and LTVs must be equipped with ESC. As ESC becomes more common in the vehicle fleet a large portion of fatal and nonfatal crashes will be prevented by this technology.

ESC systems continuously monitor several vehicle factors in order to predict impending loss of control due to excessive speed, excessive lateral acceleration, or insufficient traction. When the system predicts a loss of control it applies braking force to one or more wheels or reduces engine output to slow the vehicle. By applying brake force unequally to different wheels, ESC systems can prevent instabilities such as fishtailing and over-corrections. By reducing engine output, ESC systems can prevent unsafe levels of lateral acceleration. ESC systems are able to act quickly and discreetly, and often the driver is unaware that the system has intervened to prevent a loss of stability or control.

The evaluation methods in this report are similar to the methods of the 2007 NHTSA evaluation of ESC effectiveness that used Fatality Analysis Reporting System (FARS) data from 1997-2004 and State data from 1997-2003. This evaluation was able to include additional years of FARS data (1997-2009), allowing the inclusion of a greater variety of vehicles. For the first time in an analysis of ESC effectiveness the National Automotive Sampling System General Estimates System (NASS GES, 1997-2009) was used to create nationally representative effectiveness estimates for all police-reported crashes.

Percent effectiveness was estimated by comparing the types of crashes that vehicle models experienced immediately before and immediately after the introduction of ESC. Because optional ESC generally cannot be identified from the vehicle identification number (VIN), only models that transitioned from no available ESC system availability to standard ESC were included in this analysis. Effectiveness estimates were computed for different crash types relative to a control group of low-speed and similar crashes that are unlikely to be affected by ESC. The estimates should be interpreted as the reduction in the likelihood of a vehicle being involved in a specific type of crash as a result of ESC being added to that vehicle. Estimates marked with an asterisk are statistically significant at $p < .05$. When $p < .05$, there is 95-percent certainty that the observed results are not due to chance.

---

[1] Notice that the reduction in fatalities for cars is 23 percent and the reduction for LTVs is 20 percent, yet the combined reduction for cars and LTVs is only 18 percent. While counterintuitive, this is an accurate result that can occur when combining ratios due to changes in the mix of vehicle types over time.

The principal findings of this evaluation are the following changes in likelihood of crash involvement due to ESC:

ALL CRASHES

Crash reduction with ESC

	Cars	LTVs	Combined
Police-reported crash involvements:	5%	7%*	6%*
Fatal crash involvements:	23%*	20%*	18%*

ALL SINGLE-VEHICLE CRASHES
(except collisions with pedestrians/bicyclists/animals)

Crash reduction with ESC

	Cars	LTVs	Combined
Police-reported crash involvements:	32%	57%*	50%*
Fatal crash involvements:	55%*	50%*	49%*

ALL MULTIVEHICLE CRASHES

Crash reduction with ESC

	Cars	LTVs	Combined
Police-reported crash involvements:	-1%	0%	0%
Fatal crash involvements:	8%	8%*	6%*

FIRST-EVENT ROLLOVERS

Crash reduction with ESC

	Cars	LTVs	Combined
Police-reported crash involvements:	72%*	64%*	67%*
Fatal crash involvements:	56%*	74%*	72%*

ALL IMPACTS WITH FIXED OBJECTS

Crash reduction with ESC

	Cars	LTVs	Combined
Police-reported crash involvements:	30%	67%*	58%*
Fatal crash involvements:	47%*	45%*	39%*

SIDE IMPACTS WITH FIXED OBJECTS

Crash reduction with ESC

	Cars	LTVs	Combined
Police-reported crash involvements:	60%*	73%*	71%*
Fatal crash involvements:	65%*	65%*	58%*

CULPABLE VEHICLES IN MULTIVEHICLE CRASHES

Crash reduction with ESC

	Cars	LTVs	Combined
Police-reported crash involvements:	-3%	-1%	-1%
Fatal crash involvements:	18%	21%*	16%*

COLLISIONS WITH PEDESTRIANS/BICYCLISTS/ANIMALS

Crash reduction with ESC

	Cars	LTVs	Combined
Police-reported crash involvements:	23%	-1%	4%
Fatal crash involvements:	-9%	-11%	-9%

These results are similar to the findings of the 2007 NHTSA analysis, which due to data availability included fewer vehicle models and used State data rather than NASS GES to estimate effectiveness in all police-reported crashes. In general, the current analysis found slightly higher effectiveness in cars and slightly lower effectiveness in LTVs than the 2007 analysis. The results from the two analyses cannot be compared statistically because they share some of the same data, but the differences in any case are slight.

No significant effects were found for collisions with pedestrians, bicyclists, or animals. As in the 2007 analysis, this category continues to show small, non-significant increases in likelihood of fatal crash involvement for vehicles equipped with ESC. Because the results are non-significant, this report is not able to draw any conclusions on the effects of ESC on these types of crashes. However, NHTSA plans to keep this category on the "watch" list and repeat the analyses when more data are available. It may also be useful to examine individual cases more closely in order to explore the effects, if any, of ESC on these types of crashes.

1: Introduction

1.1   Electronic Stability Control Systems and Their Precursors

Traction control systems (TCS), also known as anti-slip regulation (ASR), were first introduced in full-sized Buick models in 1971 and later became common among high-power rear-wheel drive cars as a limited slip differential. Although it functions in a similar manner to electronic stability control (ESC), the goal of the system is different. TCS is intended to prevent wheel slippage due to excess torque, and is designed to improve acceleration and cornering performance. This is accomplished by using wheel speed sensors to detect slippage, and then reducing engine output or applying brake force when necessary.

Anti-lock brake systems (ABS) were first offered as standard equipment in 1985 on some BMW, Lincoln, and Mercedes vehicles. ABS allows brake force to be controlled at each wheel and was designed to provide optimally modulated braking force during emergency braking. Four-wheel ABS monitors each wheel for locking, and if it is detected it quickly releases the brake force to allow the wheel to resume turning. Near-optimal brake force is achieved by feathering, or quickly applying and releasing the brake many times per second. ABS prevents loss of steering due to front-wheel lockup, prevents yawing or "fish-tailing" due to rear-wheel lockup and on many surfaces reduces stopping distance relative to a skidding vehicle.[2]

ESC integrates the mechanisms and functions of ABS and TCS and adds further functionality. ESC is a computerized system that continuously monitors speed, steering wheel position, brake force at each wheel, yaw rate and lateral acceleration. This input allows the system to detect loss of control due to excessive speed, lateral acceleration or insufficient traction. When loss of control is detected, the system acts by applying braking force to one or several wheels or by reducing engine output in order to slow the vehicle or correct its path. For example, if clockwise yaw is detected the system may apply brake force to the front left wheel in order to counteract the vehicle's rotation. This action takes place so quickly that the system is essentially predictive, preventing loss of control before it occurs. Often the driver is unaware that the system has acted.

1.2 Relevant Legislation and Market Share

On April 6, 2007, the National Highway Traffic Safety Administration published final rule for Federal Motor Vehicle Safety Standard (FMVSS) No. 126 that required that passenger cars, multipurpose passenger vehicles (MPVs), and trucks and buses with a gross vehicle weight rating (GVWR) of 10,000 pounds or less be equipped with an ESC system composed that complies with the standard. The standard specified the following phase-in schedule:

---

[2] Kahane, C.J. (2004). *Lives Saved by the Federal Motor Vehicle Safety Standards and Other Vehicle Safety Technologies, 1960-2002*, (NHTSA Technical Report. DOT HS 809 833), pp.25-31. Washington, DC: National Highway Traffic Safety Administration.

Model Year	Production Beginning Date	Requirement
2009	September 1, 2008	55% with carryover credit
2010	September 1, 2009	75% with carryover credit
2011	September 1, 2010	95% with carryover credit
2012	September 1, 2011	Fully effective

ESC was first available on Mercedes and BMW vehicles in 1987. For the next several years introduction was led mostly by imports, sport utility vehicles and luxury vehicles.

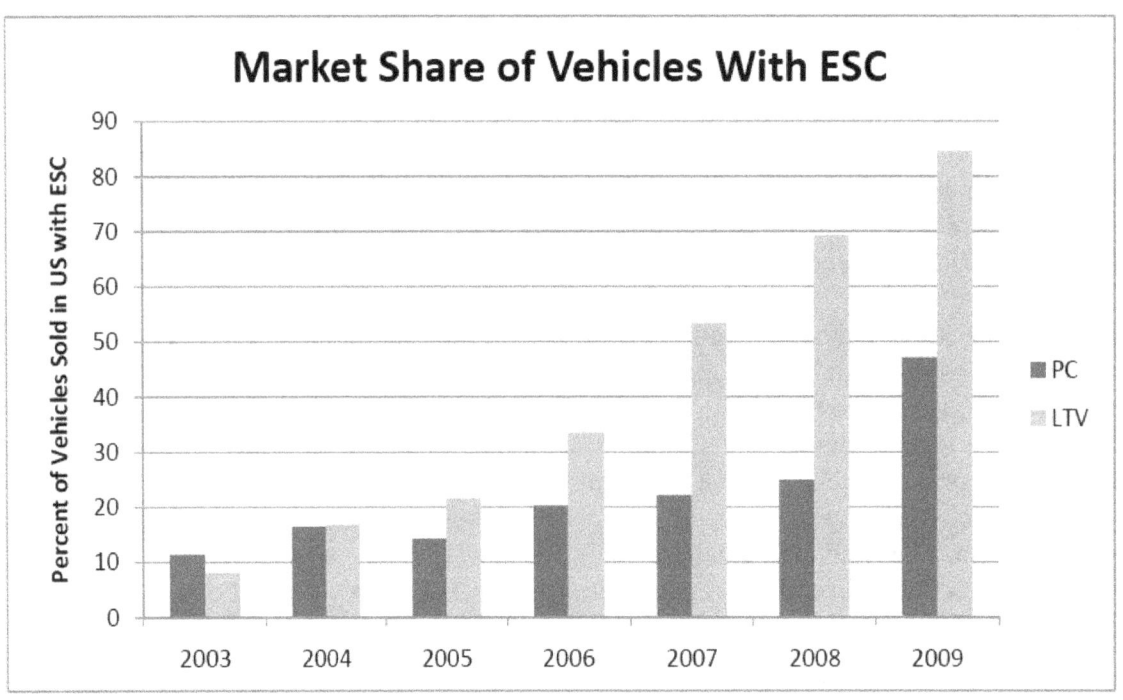

Through model year 2005, ESC was installed on less than 20 percent of the vehicles sold in the U.S. Due to mounting evidence of the effectiveness of ESC and the ensuing legislation, from 2006 on there was a sharp rise in the number of cars sold with ESC installed. Although ESC is mandated on all new vehicles of model year 2012 or later, it will take several years for ESC equipped vehicles to saturate the on-road fleet.

1.3   Previous Analyses of Effectiveness

The first study to evaluate the effectiveness of ESC was published by Aga and Okada in 2003.[3] This report analyzed crash data from three Toyota passenger car make-models and found a 36-percent reduction in single-vehicle crash rates and a 28-percent reduction in head–on collision rates attributable to ESC.

The first NHTSA evaluation of ESC effectiveness was conducted in 2004[4] and used the Fatality Analysis Reporting System data (FARS, 1997-2003) and State data (5 States, 1997-2002) to compare crash rates in vehicle models before and after ESC was made standard on those vehicles. Most of the vehicles used in this analysis were luxury sedans and SUVs since these were the first types of vehicles to offer standard ESC. The preliminary results from this study found that single-vehicle passenger car crashes were reduced by 35 percent and single-vehicle SUV crashes were reduced by 67 percent with the introduction of ESC. For fatal single-vehicle crashes, the reductions were 30 percent for passenger cars and 63 percent for SUVs.

Also in 2004, the Insurance Institute for Highway Safety (IIHS) published a study[5] that evaluated ESC effectiveness by comparing otherwise identical vehicle models with and without optional ESC systems. Based on all police-reported crashes from 7 States the report concluded that ESC reduced single-vehicle crash involvement risk by 41 percent and single-vehicle injury crash involvement risk by 41 percent as well. For fatal single-vehicle crashes, the estimated benefit was a 56-percent reduction.

In 2007, NHTSA published its most comprehensive effectiveness analysis to date.[6] It expanded on the previous NHTSA evaluation with additional years of FARS (1997-2004) and State data (1997-2003), and found similar effectiveness results. With more data, this analysis was able to investigate specific types of crashes, and found, among other large reductions, a 70-percent reduction in fatal rollover crashes in passenger cars and an 88 percent reduction in fatal rollover crashes in LTVs.   In general, LTVs showed larger crash reductions due to ESC than passenger cars, with a 28-percent overall reduction in fatalities for LTVs and a 14-percent overall reduction in fatalities for passenger cars. A small, non-significant increase in collisions with pedestrians, bicyclists, or animals was found. This analysis also compared two- and four-channel ESC systems, and found a significantly larger reduction in police-reported crashes for the four-channel systems.

In May 2010, the IIHS published an analysis[7] that compared vehicle models before and after receiving standard ESC. This analysis found a 20-percent reduction in multiple-vehicle fatal crash involvements and a 49-percent reduction in single-vehicle crash involvements attributable to ESC. Effectiveness was found to be slightly higher for LTVs than for cars, but this difference was not statistically significant. While the vehicle models included in the IIHS analysis were

---

[3] Aga, M., & Okada, A. (2003) Analysis of Vehicle Stability Control (VSC)'s Effectiveness from Accident Data, Paper Number 541, *Proceedings of the 18th International Technical Conference on the Enhanced Safety of Vehicles. Nagoya, Japan.*

[4] Dang, J. (2004) *Preliminary Results Analyzing the Effectiveness of Electronic Stability Control (ESC) Systems*, (Report No. DOT HS 809 790), Washington, DC: National Highway Traffic Safety Administration.

[5] Farmer, C. (2004). Effect of Electronic Stability Control on Automobile Crash Risk, *Traffic Injury Prevention*, Vol. 5, pp.317-325.

[6] Dang, J. (2007) *Statistical Analysis of the Effectiveness of Electronic Stability Control (ESC) Systems – Final Report,* (Report No. DOT HS 810 794). Washington, DC: National Highway Traffic Safety Administration.

[7] Farmer, C. M. (2010). Effects of Electronic Stability Control on Fatal Crash Risk. Arlington, VA: Insurance Institute for Highway Safety.

almost identical to the vehicles included in the present analysis, the methods for modeling effectiveness differed. The IIHS report used vehicle registrations to model exposure to crash risk, and limited the inclusion of model years within vehicle model only by major vehicle redesign.

1.4   Goals of the Evaluation

The primary goals of this analysis are to expand on and clarify the findings of the 2007 NHTSA analysis by using a greater variety of vehicles and several additional years of crash data. Previous research suggests that ESC has a large effect on fatality reduction and overall crash prevention. It is important to understand as clearly as possible the changes to the crash environment that will occur as a larger portion of the passenger vehicle fleet is equipped with ESC. This analysis will be better able to generalize the benefits of ESC due to the use of the National Automotive Sampling System – General Estimates System (NASS GES) to estimate the effects of ESC on all fatal and non-fatal crashes. This data is a nationally representative stratified sample of all police-reported crashes in the United States. The use of FARS data, a complete census of fatal crashes in the United States, will allow an in-depth analysis of the effects on all fatal crashes in the nation.

The principal evaluation questions are:

- What is the effect of ESC on all police-reported crashes?
- What is the effect of ESC on fatal crashes?
- What are the effects of ESC on specific types of crashes?
- How does the effectiveness of ESC differ across passenger cars and LTVs?
- What is the effect, if any, of ESC on collisions with pedestrians, bicyclists, or animals?

2: Methods

2.1   The Risk Ratio

The methodology for this evaluation is similar to that of the 2007 NHTSA evaluation. By examining the types of crashes that vehicle models are involved in immediately prior to and subsequent to the introduction of ESC, one can estimate the effectiveness of ESC by using contingency tables to compute associated risk ratios.

This method allows analysis of complex phenomena that are difficult or impossible to analyze using linear models. In this case, one wishes to determine the effect ESC has on preventing motor vehicle crashes. An intuitive way to do this might be to compare the rates of-reported crashes in vehicles with and without ESC. But to do this one must account for exposure and any other driver or vehicle characteristics that may affect these rates. From a practical standpoint this is very difficult to do with linear modeling and with the data available. This analysis addresses these problems by using crash types that are not likely to be affected by ESC as a normalizing or control factor. [8]

For example, say we are interested in the effect ESC has had on rollover crashes in a specific vehicle model. Simply comparing the number of rollovers in the two years before and two years after the introduction of ESC will not account for possible changes in the volume of sales, vehicle miles travelled, etc. The risk ratio uses the following logic: If ESC has no effect on rollovers, then the ratio of vehicle rollovers to control-group collisions unlikely to be affected by ESC (such as being struck in the rear while parked) should remain the same before and after the introduction of ESC. Any changes in the exposure rate over this time will be captured by the change in the rate of control-group collisions.

The following equation estimates ESC effectiveness on fatal rollover prevention in one of the vehicle models included in this analysis by using data from the two model years before and after introduction of ESC for this model (2004-2005 for the before ESC crashes and 2006-2007 for the after ESC crashes) from the FARS database. If ESC has no effect on the incidence of rollover crashes, then the ratio of rollover to control crashes will be similar in the time period before ESC and the time period after ESC and the risk ratio will be close to 1.000.

$$\left(\frac{\#\ of\ rollovers\ after\ ESC}{\#\ of\ control\ accidents\ after\ ESC}\right) \Big/ \left(\frac{\#\ of\ rollovers\ before\ ESC}{\#\ of\ control\ accidents\ before\ ESC}\right)$$

$$= \left(\frac{5}{38}\right) \Big/ \left(\frac{19}{87}\right)$$

risk ratio $= 0.602$

percent effectiveness $= (1 - 0.602) * 100 = 40\%$

The resulting risk ratio measures the effectiveness of ESC at reducing rollover fatalities. A risk ratio less than one implies a reduction in fatalities following introduction of ESC. When the risk ratio is subtracted from one, the result is the percent effectiveness of ESC. In this example, the effectiveness of ESC is estimated to be 40 percent. In other words, adding ESC to a vehicle of this specific model without ESC is estimated to result in a 40-percent reduction in the probability that that vehicle will be involved in a fatal rollover crash.

The data can also be arranged in a 2x2 contingency table as below. Given that the null hypothesis is that frequency of crash type (columns) is independent of ESC (rows), p-values and confidence intervals can be computed using methods appropriate to a 2x2 Chi-squared analysis, specifically Pearson's chi-squared and Cochran-Mantel-Haenszel intervals.

---

[8] Evans, L. (1986). Double Pair Comparison - A New Method to Determine How Occupant Characteristics Affect Fatality Risk in Traffic Crashes, *Accident Analysis and Prevention*, Vol. 18, pp. 217-227.

# of vehicle rollovers after ESC	# of vehicle involvements in control accidents after ESC
# of vehicle rollovers before ESC	# of vehicle involvements in control accidents before ESC

This was the method of analysis for the FARS data, which is a census of fatal crash involvements in the United States. Ordinary chi-square tests may be applied to the tables of FARS data, which are discrete counts. The GES data, which is a stratified cluster sample of crashes in the United States, requires additional steps. The data is weighted and was analyzed using SAS PROC SURVEYLOGISTIC to model the effect of ESC on crash involvement. Although the resulting estimates are risk ratios and percent effectiveness, the survey procedure is necessary to properly assess the design effects of stratification and clustering.

2.2   Control and Treatment Crash Types

The method of analysis described above requires that vehicles are classified as belonging to either the control or a treatment group based on the type of crash involvement. An ideal control group vehicle would be a stationary vehicle that is struck by another motorist since the presence or absence of ESC in this vehicle would *a priori* have no effect on the probability of crash involvement (this is not true for the striking vehicle, which is why this determination is made at the vehicle level rather than the crash level). However, there are not enough "ideal" control vehicle cases to compose an adequate control group, so vehicles are assigned to the control group if their accident involvement is deemed unlikely to have been affected by the presence of ESC. The following list describes the circumstances under which a vehicle is assigned to the control group:

- Hit while parked/stopped;
- backing/parking/low-speed (1-10 mph);
- struck in rear; or
- non-culpable involvement in a multivehicle crash on a dry road.

Non-culpable involvements on dry roads make up a large portion of the control group, and this category relies heavily on the accuracy and completeness of the accident description included in the data files. To test if this group of crashes is introducing any bias, the NASS GES estimates were recomputed without the non-culpable involvements on dry roads in the control group. Reassuringly, the resulting weighted estimates were almost identical to those computed when they were included. Since the treatment group will be broken down by more specific crash types, each of which will be compared to the same control group, an accurate and uncompromised group of control vehicles is an important component of the analysis.

All vehicles that are not classified as control group vehicles are eligible to be included in a treatment crash group. The treatment groups are defined using available data gathered from sources such as the police accident report, which specifies the circumstances of the crash and the role of each vehicle involved and include the following:

All non-control-group vehicles: This group includes all of the vehicles in the data files that do not meet the criteria for the control group. There will be a variety of crashes in this group, and it is not expected to show as large of an effect of ESC as some of the other treatment groups that are specifically chosen because they are likely to be affected by vehicle control and stability.

All single-vehicle crashes (except collisions with pedestrians/bicyclists/animals): This group includes all single-vehicle crashes in the data files, except for those involving pedestrians, bicyclists, or animals, which are analyzed separately. Research has shown that ESC is particularly effective in preventing single-vehicle types of crashes, which are very likely to be the result of loss of vehicle control.

First-event rollovers: This group is a subset of the single-vehicle crashes and is defined by examining the first harmful event in the crash sequence recorded in the data files. Subsequent-event rollovers, such as a vehicle that strikes a fixed object and rolls as a result, are not included in this group.

All impacts with fixed objects: This group is a subset of single-vehicle crashes and includes all single-vehicle run-off-road crashes except first-event rollovers, collisions with pedestrians, bicyclists, animals, or other movable objects such as trains, and non-collisions such as immersion in water or a person falling off a moving vehicle.

Side impacts with fixed objects: This group is a subset of all impacts with fixed objects. These vehicles are analyzed separately because side impacts are particularly characteristic of loss of vehicle control.

Culpable vehicles in multivehicle crashes: This group consists of vehicles that have been identified as the culpable party in a multivehicle crash. This group may contain vehicles that experienced loss of control, but may also contain vehicles that were involved in crashes that would not have been affected by ESC. In past analyses these vehicles have shown a smaller benefit from ESC than those involved in single-vehicle crashes.

Collisions with pedestrians/bicyclists/animals: This group is singled out for analysis because the 2007 NHTSA analysis showed a small, non-significant increase in crash risk for vehicles with ESC. One way ESC functions is by attenuating driver steering and/or braking input that may result in loss of control, and it is possible that this could contribute to a reduction in the ability to make emergency evasive maneuvers.

The effectiveness of ESC in other populations of interest can be derived using results from the preceding groups. The two derived estimates in this report are effectiveness in all crashes and effectiveness in all multivehicle crashes. These estimates must be derived rather than computed because the control group includes members of these populations of interest.

All Crashes: This estimate is derived using results from the all non-control group crashes. Because an assumption of the analysis is that ESC will have no effect on the control group crashes we can estimate effectiveness in all crashes with the following formula:

$$effectiveness = \theta_t * x_t / (x_t + y_c)$$

Where

θ_t = the estimated effectiveness for non-control group crashes
x_t = the number of non-control group crashes before ESC
y_c = the number of control group crashes before ESC

Because all crashes in the data are either contained in the control or non-control group, this will give an estimate of effectiveness in all crashes. The confidence interval for this estimate can be derived by replacing θ_t with the upper and lower bound estimates for the 95 percent confidence interval of the estimated effectiveness for all non-control group crashes.

All Multivehicle Crashes: This estimate can be derived from the results of the culpable vehicles in multivehicle crashes group using the same logic and formula as the "All Crashes" group above, with
x_t = the number of culpable vehicles in multivehicle crashes
θ_t = the estimated effectiveness for culpable vehicles in multivehicle crashes

Because all of the control group crashes are multivehicle crashes, and because all multivehicle crashes in the data files are contained in either the control group or in the culpable vehicles in multivehicle crashes group, this will give an estimate of effectiveness in all multivehicle crashes.

2.3   Selection of Vehicle Models

ESC is often offered as an optional feature whose presence is impossible to determine from the VIN. Accordingly, only vehicle models that transitioned from no ESC to standard ESC could be included in the analysis. Eligible vehicles were identified using previous NHTSA analyses, www.safercar.gov, and information provided by vehicle manufacturers. The two model years before and the two model years after the introduction of ESC were included when possible. In cases where a major vehicle redesign took place during this period, the included model years were truncated to ensure that only similar vehicles were compared. In some of the more recent models, rollover sensors were introduced and present a potential confound for analyses of rollover crashes. For these vehicles, model years were truncated in analyses including rollover crashes so that the presence of rollover sensors was consistent across all included model years. Some vehicle models are included that had a period of time that ESC was offered as an option; these optional model years are removed. The following tables list the included vehicles and model years. Table 1 contains the vehicles carried over from the 2007 NHTSA analysis, and Table 2 lists the vehicles new to this analysis.

Table 1: Vehicle Models Carried Over From 2004 Analysis			
Make	Model	Years Before ESC Included	Years After ESC Included
Acura	RL	1998-1999	2000-2001
Acura	MDX	2001-2002	2003-3004
Audi	A6	2000*	2001-2002
Audi	TT	2000*	2001-2002
BMW	300	1998-1999	2000-2001
BMW	528i	1998	2000-2001
BMW	540i	1997-1998	1999-2000
BMW	740i	1997*	1998-1999
BMW	740iL	1997*	1998-1999
BMW	Z3	1998-1999	2000-2001
Buick	Park Avenue Ultra	1998-1999	2000-2001
Cadillac	Deville Concours	1997*	1998-1999
Cadillac	Escalade	2002*	2003-2004
Cadillac	Seville	1998-1999	2000*
Chevrolet	Express 3500	2002-2003	2004-2005
GMC	Yukon Denali	2001-2002	2003-3004
GMC	Yukon Denali XL	2001-2002	2003-3004
GMC	Savana G3500 EXT RWD	2002-2003	2004-2005
Lexus	GS-400	1998*	1999-2000
Lexus	LS-400	1997-1998	1999-2000
Lexus	LX470	1999*	2000-2001
Lexus	RX300	1999-2000	2001-2002
Mercedes	ML 320	1998	1999-2000
Mitsubishi	Montero Limited	2001-2002	2003-3004
Toyota	Land Cruiser	1999-2000	2001-2002
Toyota	RAV4	2002-2003	2004-2005
Toyota	4Runner	1999-2000	2001-2002
Volkswagen	Passat GLX	2002	2004
* Period truncated due to redesign			

Not all vehicles in the 2007 NHTSA evaluation are included in this analysis. Previously, if "similar" vehicle models were available for before and after ESC conditions these vehicles were included in the analysis. For example, the 2002 Volkswagen Beetle was included in the "before ESC" group and the 2002-2004 Volkswagen Beetle Turbo S was included in the "after ESC" group.

Table 2 below lists the new vehicles added to the analysis. Some of these vehicles had a single year of data removed from all analyses including rollover crashes because of the addition of rollover sensors. These years are listed in the last column of the table.

Table 2: New Vehicles				
Make	Model	Years Before ESC Included	Years After ESC Included	Rollover Sensor Year Removed
Buick	Rainier	2004-2005	2006-2007	2004
Chevrolet	Avalanche	2004-2005	2006-2007	
Chevrolet	Trailblazer	2004-2005	2006-2007	2004
Chevrolet	Equinox	2005-2006	2007-2008	
Chrysler	Pacifica	2005-2006	2007-2008	
Dodge	Durango	2005-2006	2007-2008	
Dodge	Sprinter	2005-2006	2007-2008	
Ford	Explorer	2003-2004	2005*	
GMC	Envoy	2004-2005	2006-2007	2004
Honda	CR-V	2003-2004	2005-2006	
Honda	Odyssey	2003-2004	2005-2006	
Honda	S2000	2004-2005	2006-2007	
Honda	Civic Si	2006*	2007-2008	
Honda	Element	2005-2006	2007-2008	
Isuzu	Ascender	2004-2005	2006-2007	2004
Jeep	Grand Cherokee	2004-2005	2006-2007	
Jeep	Liberty	2004-2005	2006-2007	
Kia	Sorento	2005-2006	2007-2008	
Mercedes	SLK	1999-2000	20001-2002	
Mercury	Mariner	2006-2007	2008-2009	
Mercury	Mountaineer	2003-2004	2005*	
Pontiac	Torrent	2005-2006	2007-2008	
Subaru	Impreza WRX	2006-2007	2008-2009	
Suzuki	Grand Vitara	2004-2005	2006-2007	
Volkswagon	Eurovan	1999-2000	2001-2002	
* Period truncated due to redesign				

The new vehicles added to this analysis are more diverse than the early adopters in the 2007 analysis, which were mostly luxury cars and LTVs. The inclusion of these vehicles will help to make the resulting effectiveness estimates more representative of the current vehicle fleet.

3: Effect of ESC in All Police-Reported Crashes

3.0   Summary

NASS GES data files from 1997 to 2009 were used to estimate the effectiveness of preventing vehicle involvement in treatment group crashes of any severity. This data is compiled annually from a nationally representative probability sample of every police-reported crash in the U.S. Although many crashes are not reported to police, unreported crashes are unlikely to involve significant personal injury or major property damage.

The results show large reductions in loss-of-control crashes for vehicles equipped with ESC. There was a 50-percent reduction in all single-vehicle crashes (excluding collisions with pedestrians, bicyclists, or animals), a 71-percent reduction in side impacts with fixed objects, a 58-percent reduction in all impacts with fixed objects, and a 67-percent reduction in first-event rollovers for vehicles with ESC. All of these reductions were statistically significant.

When passenger cars were analyzed separately, the only significant results were a 60-percent reduction in side impacts with fixed objects and a 72-percent reduction in first-event rollovers. Other crash types showed large reductions but were not statistically significant. This is likely a result of a small sample size as most of the vehicles eligible for analysis were LTVs.

LTVs showed large reductions in all of the single-vehicle crash categories, all of which were statistically significant. ESC appeared to have no effect on culpable parties in multivehicle crashes or on collisions with pedestrians, bicyclists, or animals.

It is difficult to compare these results to previous analyses because this is the first time that this data has been used in an analysis of ESC effectiveness. However, the results are similar to the estimates in the 2007 NHTSA analysis that were derived from State data, which also includes all police-reported crashes.

3.1   Results for All Vehicles

There are 8,040 total NASS GES cases included in this analysis taken from fourteen years of crash data files. Vehicles were identified as described in the methods section and then placed in a database subset. NASS GES data is available at three different levels, the crash level, the vehicle level, and the occupant level. Crash types were assigned to each vehicle case in the vehicle level data using variables at the crash and vehicle levels. Analysis was conducted using SAS PROC SURVEYLOGISTIC to properly specify the survey design.

The results for each analyzed crash category are given in Table 3 below. The reported statistics are unweighted and weighted risk ratios (see section 2.1 for an explanation of how risk ratios are computed from the crash data), 95-percent confidence intervals for the weighted risk ratios, and percent effectiveness derived from the weighted risk ratio [(1 − risk ratio) * 100]. Any estimate with a 95-percent confidence interval whose upper and lower bounds are both less than 1.000 is statistically significant at the $p < .05$ level and is marked with an asterisk.

The unweighted risk ratio estimates are not nationally representative, but are reported because they can lend insight into the reliability of the weighted estimates. Large differences between estimates based on the unweighted and weighted data are often a symptom of insufficient sample size. This does not appear to be a problem with this data, as weighted and unweighted estimates do not differ substantially.

Table 3: ESC Effectiveness in All Police-Reported Crashes (NASS GES) All Vehicles				
Crash Type	Risk Ratio (Unweighted)	Risk Ratio (Weighted)	95% CI (Wald)	Weighted % Effectiveness
All crashes[†]	.917	.937	(.902, .976)	6%*
All non-control group	.813	.845	(.759, .941)	16%*
All single-vehicle (except ped/bikes/animals)	.514	.496	(.423, .581)	50%*
first-event rollovers	.295	.332	(.223, .494)	67%*
All impacts w/ fixed obj.	.513	.424	(.342, .525)	58%*
Side impacts w/ fixed obj.	.372	.29	(.187, .449)	71%*
All multivehicle[†]	.979	1.003	(.974, 1.035)	0%
Culpable multivehicle	.924	1.011	(.901, 1.134)	-1%
Peds/Bikes/Animals	1.057	.955	(.681, 1.340)	4%
* = statistically significant at p < .05				
†= derived estimate, see section 2.2				

The estimates of effectiveness in all crashes and in all multivehicle crashes are derived using the method described in section 2.2. To illustrate, the estimate for effectiveness in all crashes was derived from the estimate for all non-control group vehicles. Without ESC, there were 577,007 weighted control group involvements and 395,905 weighted non-control group involvements. The point estimate for ESC effectiveness in the non-control group involvements is 15.5 percent ([1-.845]*100), and the reduction in all crashes is estimated using the following formula: [395,905/(577,007 + 395,905)] * 15.5 = 6.3-percent reduction in all crashes.

The 95-percent confidence bounds, derived using bounds for the original estimate, are from 2.4 percent to 9.8 percent. The 2007 analysis estimated an 8-percent reduction, slightly larger than the estimate found here but within the current error bounds.

All of the single-vehicle crash categories showed large significant decreases in crash risk for ESC-equipped vehicles. These decreases were particularly large for the crash types hypothesized to be affected most by vehicle control and stability: first-event rollovers (67% reduction) and side impacts with fixed objects (71% reduction). The results for multivehicle crashes and for collisions with pedestrians, bicyclists, or animals are less clear. These estimates are close to zero effect and have large confidence intervals.

3.2   Results by Vehicle Type

Given that light trucks and vans are on average more prone to loss of control and rollover crashes than passenger cars, both types of vehicles are analyzed separately in this section and the results are given in the following tables. Although the types of vehicles included in this analysis are more equally balanced between cars and LTVs than in previous studies, the bulk of the data still comes from LTVs, as these vehicles were much more likely to be equipped with ESC during the time period included in the study (see market-share graph in section 1.2). This will result in less precise estimates and larger confidence intervals for the passenger car data.

When passenger cars were analyzed separately, there were too few cases to obtain significant estimates for most of the crash types. The only significant results were a 60-percent reduction in side impacts with fixed objects and a 72-percent reduction in first-event rollovers. Because small sample sizes lead to large confidence intervals, only very large estimates will be statistically significant. The point estimates for other crash types, while not significant, are similar to the combined results.

Table 4: ESC Effectiveness in All Police-Reported Crashes (NASS GES) Passenger Cars Only				
Crash Type (PC Only)	Risk Ratio (Unweighted)	Risk Ratio (Weighted)	95% CI (Wald)	Weighted % Effectiveness
All crashes[†]	.92	.952	(.865, 1.067)	5%
All non-control group	.825	.881	(.666, 1.166)	12%
All single-vehicle (except ped/bikes/animals)	.652	.677	(.452, 1.013)	32%
first-event rollovers	.347	.278	(.090, .857)	72%*
All impacts w/ fixed obj.	.683	.696	(.451, 1.074)	30%
Side impacts w/ fixed obj.	.467	.397	(.192, .818)	60%*
All multivehicle[†]	.966	1.008	(.936, 1.104)	-1%
Culpable multivehicle	.884	1.028	(.750, 1.407)	-3%
Peds/Bikes/Animals	.990	.767	(.456, 1.291)	23%
* = statistically significant at p < .05				
†= derived estimate, see section 2.2				

LTVs have a larger sample size and when analyzed separately from passenger cars the estimated reductions in loss-of-control crashes were all large and significant. The only non-significant crash types were collisions with pedestrians, bicyclists or animals and multivehicle crashes.

13

Table 5: ESC Effectiveness in All Police-Reported Crashes (NASS GES) Light Trucks and Vans Only				
Crash Type (LTV Only)	**Risk Ratio (Unweighted)**	**Risk Ratio (Weighted)**	**95% CI (Wald)**	**Weighted % Effectiveness**
All crashes[†]	.912	**.933**	(.898, .972)	7%*
All non-control group	.800	**.836**	(.750, .932)	16%*
All single-vehicle (except ped/bikes/animals)	.455	**.432**	(.364, .512)	57%*
first-event rollovers	.309	**.359**	(.239, .541)	64%*
All impacts w/ fixed obj.	.436	**.332**	(.263, .418)	67%*
Side impacts w/ fixed obj.	.332	**.268**	(.162, .445)	73%*
All multivehicle[†]	.979	**1.003**	(.972, 1.037)	0%
Culpable multivehicle	.923	**1.011**	(.894, 1.143)	-1%
Peds/Bikes/Animals	1.075	**1.013**	(.709, 1.013)	-1%
* = statistically significant at p < .05				
†= derived estimate, see section 2.2				

There are a couple of interesting observations to be made about the results for PCs and LTVs. In the 2007 NHTSA analysis, LTVs showed much larger effectiveness estimates than passenger cars. In this analysis the results seem much more similar across vehicle type. There could be a variety of reasons for this, such as improved stability in later models of LTV's, inclusion of more compact utility vehicles (CUVs) in the LTV group, inclusion of more non-luxury models of passenger cars, and others. More detailed analysis did not reveal any one specific cause for the increased similarity of effectiveness across cars and LTVs.

Some further analysis conducted on the LTV sample seemed to suggest that there is a unidirectional time effect for ESC effectiveness. The newer LTV models in general showed smaller effectiveness estimates than the older models, although these differences were not statistically significant. The 2010 IIHS analysis using similar vehicles included an analysis of effectiveness by vehicle make. While the sample sizes were too small to yield statistically significant results, there did seem to be a trend for makes that had introduced ESC earlier to have higher effectiveness estimates. This is likely a worthwhile avenue of future research; if newer vehicles show smaller benefits due to an interaction with other improvements in stability and handling, then this effect should be considered when estimating the effects that ESC will have when it is added to new vehicles.

4: Effect of ESC in Fatal Crashes

4.0   Summary

To estimate the effect of ESC on fatal crash involvement, data from the Fatal Accident Reporting System (FARS) was used to compare relative fatality rates in vehicle models before and after introduction of standard ESC. FARS is a census of every fatal crash in the Nation. Data is collected at the State level using several sources such as police accident reports, driver licensing files, and vehicle registration files. The data is standardized across States and combined into a single database that NHTSA releases yearly.

The analysis of fatal crashes found a significant 18-percent reduction in the likelihood that a vehicle would be involved in any type of fatal crash and showed large, statistically significant reductions in fatalities for several different types of crashes for vehicles equipped with ESC. When passenger cars and light trucks and vans were analyzed separately, the results were very similar across vehicle type.

The results generally agree with previous estimates of ESC effectiveness. As in the analysis of all police-reported crashes, the analysis of fatal crashes found effectiveness to be similar across cars and LTVs, whereas in earlier reports effectiveness was notably higher in LTVs than in cars. When compared to the 2007 NHTSA analysis, which used similar crash categories, it appears as though the effectiveness in cars has increased slightly and the effectiveness in LTV's has decreased slightly from previous estimates.

4.1   Fatal Crash Results for All Vehicles

The effect of ESC on fatal crashes was estimated using data in 1997-2009 FARS. The same vehicle models that were used in the NASS GES analysis were used here as well. This analysis included 6,172 vehicle cases from the FARS database.

The following table presents the counts of vehicle cases, risk ratios, 95-percent confidence intervals for the risk ratios, and percent effectiveness estimates for each crash category. The confidence intervals were computed with SAS PROC FREQ, which uses the Cochran-Mantel-Haenszel method of interval construction.

Table 6: ESC Effectiveness in Fatal Crashes (FARS) All Vehicles					
Crash Type	Vehicles w/o ESC	Vehicles w/ ESC	Risk Ratio	95% CI (CMH)	% Effectiveness
Count of control crashes	1477	787			
All crashes[†]	4296	1876	.82	(.77, .875)	18%*
All non-control group	2819	1089	.725	(.649, .81)	27%*
All single-vehicle (except ped/bikes/animal)	1294	348	.505	(.436, .584)	49%*
first-event rollovers	502	76	.284	(.22, .367)	72%*
All impacts w/ fixed obj.	648	212	.614	(.514, .733)	39%*
Side impacts w/ fixed obj.	152	34	.42	(.287, .615)	58%*
All multivehicle[†]	2384	1192	.939	(.895, .988)	6%*
Culpable multivehicle	907	405	.84	(.725, .969)	16%*
Peds/Bikes/Animals	415	242	1.094	(.914, 1.311)	-9%
* = statistically significant at $p < .05$					
†= derived estimate, see section 2.2					

Estimates of ESC effectiveness at preventing fatal single-vehicle crashes (excluding collisions with animals, bicycles, or pedestrians) are very similar to the results of the analysis of all police-reported crashes. This is not a surprising result, since single-vehicle crashes are likely to be loss-of-control crashes that occur at high speeds, regardless of whether they are fatal or not. In other words, one would expect fatal single-vehicle crashes to be fairly representative of single-vehicle crashes in general.

For the single-vehicle crashes the results are clear; ESC is highly effective at preventing fatalities from these types of crashes. These estimates also show narrow confidence intervals, indicating small variance and low volatility. The reduction for all non-control crashes (27%) is also impressively large considering the variety of crashes included in this category.

These results are very similar to the effectiveness estimates for the same crash types reported in the 2007 NHTSA evaluation using FARS data from 1997-2004, however a detailed comparison will not be given since slight changes in vehicle inclusion criteria and statistical methods make direct contrasts inappropriate.

Although there are some differences between the FARS and NASS GES derived estimates, most notably in the effects on collisions with pedestrians, bicyclists or animals and the culpable involvements in multivehicle crashes, these estimates are not statistically significant regardless of data source and interpretation of ESC effectiveness in these types of crashes will be deferred until sufficient data is available.

4.2   Fatal Crash Results by Vehicle Type

The results for passenger cars are very similar to the overall results. Although the reduction in culpable vehicles in multivehicle accidents is slightly larger, the sample size is smaller and the reduction is still non-significant. Despite the reduced sample size, the single-vehicle crash categories and the all crashes and non-control group give statistically significant estimates of crash reduction.

Table 7: ESC Effectiveness in Fatal Crashes (FARS) Passenger Cars Only					
Crash Type (PCs)	Vehicles w/o ESC	Vehicles w/ ESC	Risk Ratio	95% CI (CMH)	% Effectiveness
Count of control crashes	177	174			
All crashes[†]	656	495	.768	(.657, .911)	23%*
All non-control group	479	321	.682	(.53, .878)	32%*
All single-vehicle (except ped/bikes/animal)	253	125	.503	(.373, .678)	50%*
first-event rollovers	49	21	.436	(.251, .757)	56%*
All impacts w/ fixed obj.	170	89	.533	(.387, .742)	47%*
Side impacts w/ fixed obj.	49	17	.353	(.196, .637)	65%*
All multivehicle[†]	333	299	.916	(.81, 1.055)	8%
Culpable multivehicle	156	125	.815	(.595, 1.117)	18%
Peds/Bikes/Animals	45	48	1.085	(.687, 1.714)	-9%
Total number of cases	656	495			
* = statistically significant at p < .05					
†= derived estimate, see section 2.2					

The analysis of passenger car involvements by crash type shows large reductions across crash types, consistent with previous effectiveness analyses. The 23 percent effectiveness estimate for all crashes suggests that nearly a quarter of all fatal crashes in passenger cars may be prevented by adding ESC.

Table 8: ESC Effectiveness in Fatal Crashes (FARS) Light Trucks and Vans Only					
Crash Type (LTVs)	Vehicles w/o ESC	Vehicles w/ ESC	Risk Ratio	95% CI (CMH)	% Effectiveness
Count of control crashes	1300	613			
All crashes[†]	3640	1381	.805	(.752, .865)	20%*
All non-control group	2340	768	.696	(.614, .79)	30%*
All single-vehicle (except ped/bikes/animal)	1041	223	.454	(.382, .54)	55%*
first-event rollovers	453	55	.258	(.192, .346)	74%*
All impacts w/ fixed obj.	478	123	.546	(.438, .68)	45%*
Side impacts w/ fixed obj.	103	17	.35	(.208, .59)	65%*
All multivehicle[†]	2051	893	.923	(.888, .976)	8%*
Culpable multivehicle	751	280	.791	(.669, .935)	21%*
Peds/Bikes/Animals	370	194	1.112	(.912, 1.356)	-11%
Total number of cases	3640	1381			
* = statistically significant at p < .05					
†= derived estimate, see section 2.2					

LTV's also show large significant reductions in fatalities. This is the only analysis that showed a significant reduction in culpable multivehicle crashes. The only crash category that did not show a significant reduction was collisions with pedestrians, bicyclists, or animals, which showed an 11-percent non-significant increase.

5: Discussion

5.1   Summary

In many ways, ESC is an ideal crash avoidance technology. Because it acts so quickly and without driver input it can prevent a crash without the driver of the vehicle being aware that the system has intervened. It shares many components with the ABS system, reducing the cost for implementation. And most importantly, it has been shown by this analysis and several others, using a variety of methods, to be highly effective at preventing loss-of-control crashes.

By using NASS GES data this evaluation is able to compute nationally representative estimates of ESC effectiveness on crash involvements. This will be a valuable tool in attempts to predict the broad economic and safety related effects that ESC will have in the future.

FARS data is the most comprehensive and accurate fatal-crash data available, and this evaluation and others have shown that this data supports the assertion that ESC has a major impact on vehicle safety. The estimates derived from this data suggest that the mandatory inclusion of ESC on all new vehicles beginning in MY 2012 will save thousands of lives every year due to prevention of fatal loss of control crashes.

Although the results of this analysis and others are very encouraging, it is important to consider any possible disbenefits associated with ESC. There have been no statistically significant increases in any crash type associated with the introduction of ESC. However, small, non-significant increases in the incidence of collisions with pedestrians, bicyclists and animals were observed in this study (FARS only, not GES) and in the 2007 NHTSA evaluation. Because these effects seem to be very small, if they do indeed exist, there is not yet enough data for statistically meaningful results. While this report draws no conclusions about pedestrian crashes, NHTSA plans to keep this category on the "watch" list and repeat the analyses when more data are available. It may also be useful to examine individual cases more closely in order to explore the effects, if any, of ESC on these types of crashes.

5.2   Limitations

As discussed in section 2.1, the statistical methods applied in this report involve comparing earlier versions of vehicle models without ESC to later versions of the same model with ESC. This requires the assumption that over the included time period (two years before introduction of ESC and two years after introduction of ESC) the vehicle remains much the same aside from the introduction of ESC. Unfortunately, there is some evidence that this is not always the case. When ESC effectiveness is compared across vehicle model, it appears as though there is a tendency for effectiveness to decrease over time. It is impossible to determine whether this is due to differences in early and late adopters of ESC or if the relative frequency of treatment (crashes likely to be prevented by ESC) and control (crashes not likely to be prevented by ESC) crashes is changing over time. If there is a relative decrease in treatment crashes over time within vehicle models, then the method of analysis used in this report would lead to an overestimate of the effectiveness of ESC because these decreases would be attributed to ESC instead.

To test whether there was a decrease in treatment crashes relative to control crashes over time that was not due to the introduction of ESC, a supplemental analysis was conducted on the vehicles included in this analysis. The time periods for the vehicle model years in the study were shifted back two years, so that instead of comparing the two model years before ESC to the two model years after ESC, the third and fourth model years before ESC are compared to the first and second model years before ESC. If the assumptions are correct then this analysis should result in a risk ratio that is not significantly different from 1.000, since both time periods include only vehicles without ESC. In fact, however, the resulting risk ratios for several vehicles were significantly smaller than one (even after taking major vehicle redesigns into account).

This could be due to changes in the vehicle other than ESC during this time period that reduced the relative incidence of treatment crashes. An examination of the changes in the vehicles with particularly large reductions did not suggest any obvious candidates. However the vehicles that showed a significant decrease were notably all LTVs, and it is likely that many small changes in these vehicles resulted in greater stability over time. These results were not used to adjust the estimates in this report. It would not be valid to assume that these effects would be constant over time, i.e. that the relative reduction in non-control group crashes due to vehicle changes that took place over the four model years before the introduction of ESC will equal the relative reduction due to vehicle changes other than ESC in the four model years surrounding introduction of ESC. Because it is impossible to quantify the effects of changes other than ESC that took place at the same time as the introduction of ESC, it is impossible to gauge the magnitude of the resulting bias to the estimates in this report.

Some possible sources of this bias could be specifically identified as having potential to affect the estimates in this analysis. These were changes to vehicles that could make later versions of the same vehicle models less prone to treatment crashes than their earlier counterparts, and could therefore lead to an overestimation of the effectiveness of ESC.

1)   Vehicle Redesign:
Vehicle models change frequently, and it is important to distinguish between cosmetic changes that are unlikely to affect crashworthiness and more profound changes that may affect the vehicle's safety. In this analysis, vehicle models that changed wheelbase during the included time period were excluded if the redesign occurred in the same year as the introduction of ESC. If the redesign occurred one year before or one year after ESC was introduced, then the analyzed time period was truncated for that vehicle so that only vehicles with the same wheelbase were compared.

2)   Static Stability Factor:
Static stability factor (SSF) is a simple measurement of a vehicle's resistance to tipping and rollover. It is a measurement of how top-heavy a vehicle is and is calculated using the formula:

$$SSF = \frac{T}{2H}$$

where

T = track width; distance between the centers of the right and left tires along the axle
H = height of the center of gravity

The lower the SSF, the more likely a vehicle is to roll in a tripped single-vehicle crash. Across all passenger vehicles, SSF values tend to range from 1.00 to 1.50.[9]

The static stability factors for the included vehicles were graphed over the included time period so that large changes in the SSF could be identified using Chow's F-statistic.[10] No vehicles were identified by this method as undergoing a significant change in the SSF during the time period of interest. Although SSF tends to increase over time, the within model time periods are so brief (4 years maximum) that large changes are unlikely.

3)   Rollover Sensors:

Rollover sensors are designed to measure the lateral and vertical acceleration, speed and roll rate of a vehicle in order to predict an impending rollover. When the sensors detect an impending rollover, the control module triggers the side curtain air bags to help protect passengers against severe injury. The coincidental introduction of rollover sensors and ESC could result in safety benefits from the sensors being attributed instead to ESC.

No vehicles in the analysis received side curtain airbags in the same model year as ESC, but several GM models received sensors one year before the introduction of standard ESC. For these vehicles, all rollover crashes that occurred in the year before rollover sensors were included were excluded from analysis.

Even after addressing the sources of bias listed above, some vehicle models still showed a relative decrease in treatment crashes during the four years before the introduction of ESC. While this suggests that there may be an overestimation bias to the estimates in this report, it is not proof that such bias does in fact exist. Because steps were taken to identify and remove bias such as limiting the included within model years to two years before and two years after ESC, these estimates should be more accurate and robust than any previous effectiveness estimates.

---

[9] Walz, M. (2005). *Trends in the Static Stability Factor of Passenger Cars, Light Trucks, and Vans*. (Report No. DOT HS 809 868), Washington, DC: National Highway Traffic Safety Administration.

[10] Baltagi, B. (2008) *Econometrics*. Syracuse, NY: Springer

DOT HS 811 486
June 2011